How to THINK METRIC

W. W. Bates   Olive Fullerton

FEARON·PITMAN PUBLISHERS, INC.
Belmont, California

Fearon·Pitman Publishers, Inc.
1977 printing

© Copp Clark Publishing 1974

All rights reserved. No part of the material covered by this copyright may be reproduced in any form or by any means (whether electronic, mechanical or photographic) for storage in retrieval systems, tapes, discs or for making multiple copies without the written permission of Copp Clark Publishing.

ISBN 0-8224-3763-5
Library of Congress Catalog Card Number: 74-75325

Design/Peter Maher
Illustrations/Graham Pilsworth
Diagrams/Frank Zsigo

Contents

How To Use This Book 1

Introduction 2

A Decimal System 3

Linear Measurement 4
The Metre 4
Practice the Metre 6
The Centimetre 7
Practice the Centimetre 10
The Kilometre 11
Practice the Kilometre 13
The Millimetre 18
Practice the Centimetre and the Millimetre 19

Mass 21
The Kilogram 21
Practice the Kilogram 23
The Gram 26
Practice the Gram 28

Capacity 29
The Litre 29
The Millilitre 32
Practice the Litre and the Millilitre 34

Area 35
The Square Centimetre 35
Practice the Square Centimetre 39
The Square Metre 40
Practice the Square Metre 42
The Hectare 43

Volume   45

The Cubic Centimetre   45
Practice the Cubic Centimetre   47
The Cubic Metre   49
Practice the Cubic Metre   51

Temperature   52

The Celsius Thermometer   52
The Clinical Thermometer   54

Consumer Goods   55

Recall of Decimals   57

Quiz   57
Operations with Decimals   59
Multiplying and Dividing by 10, 100 and 1,000   61
Practice Exercise   62

Charts and Graphs   63

Conversion Tables   70

Answers   72

How To Use This Book

Working through this book will be much easier than you think. The more you do, the easier it will become. It's a good idea to work with someone, to talk about the ideas and to make the measurements together. Use the tables for recording; write in pencil and have an eraser handy.

Don't hesitate to estimate. Your first few estimates may not be close to the actual measurements, but don't worry. You will soon become an expert!

Read the table of contents to see what ideas we'll be looking at. Don't stop there, though! Be the first on your block to *think metric!*

Introduction

More than 90 percent of the world uses the metric system. In fact, the United States and Canada are the only two major countries still using the imperial system. Even England, where inches, feet and yards originated, has gone metric. With the increase in trade and in world travel, it doesn't make sense for North Americans to use a different — and often very confusing — system of measurement. So the United States is going metric too! Among other things, this means that children will learn the metric system at school . . . so parents will want to be prepared.

This little book is for anyone who is prepared to discard an obsolete system of measurement and adopt one that is logical and much easier, and that will be in use throughout the world in the near future. It will help you to understand the units you will commonly use — no others. You will get an idea of the size of these units, you will practice estimating the size of objects with them, and then you will measure accurately with them. We will *not* make any comparisons with — or conversions from — imperial units, since learning by converting is like trying to speak French by thinking in English. It won't work. At the end of the book, we do include some tables showing the relationship between the two systems. These tables may be convenient for some purposes: you may wish to metricate a favorite old recipe or compare the price of an item measured in the metric system with the same item measured in the imperial system. Otherwise, we hope you will declare the tables out of bounds.

Thinking metric is the only successful method for school children or for adults. We admit that familiarity with the imperial system will be a handicap, but if you keep trying, difficulties will disappear and you will soon be thinking metric.

A Decimal System

The metric system, like our money system, is a decimal system — it involves only multiples of 10. You are familiar with relationships such as 12 inches in a foot, 3 feet in a yard, 5½ yards in a rod. When you think about it, these numbers are really quite ridiculous. In contrast, metric relationships involve 10, 100, 1000, and so on. We can convert from one unit to another by simply shifting the decimal point. For those who need a review, working with decimals is dealt with in detail later in the book. In the meantime let's begin to think metric.

4 *Linear Measurement*

The Metre

You will need a metric tape, and, although you can get along without one, you may wish to have a metre stick as well. If you don't want to buy a metre stick, you can make one in a few minutes. Find a straight, smooth stick. The handle of an old hockey stick, a broom handle or a piece of molding will do. Measure off a metre and there you are. You could even calibrate it in centimetres after you have worked through that section of this book.

Examine the tape or stick. The metre is the base unit of length. You will express many lengths in metres; for example, a room may be five metres long and four metres wide.

To get a feeling for the length of a metre, hold the stretched metre tape, or the stick, in front of you. Look around the room at familiar objects. Think about one object at a time. Is the window sill about the same length as the metre tape? Is it longer or shorter? Now check how well you estimated by measuring. With a little practice, you will find it easy to estimate in metres.

When writing metric units, symbols are used. The symbol m is used for the metre. Notice that lower case letters are used and there are no periods. Symbols are not treated as abbreviations nor are they pluralized.

A Metre Chart

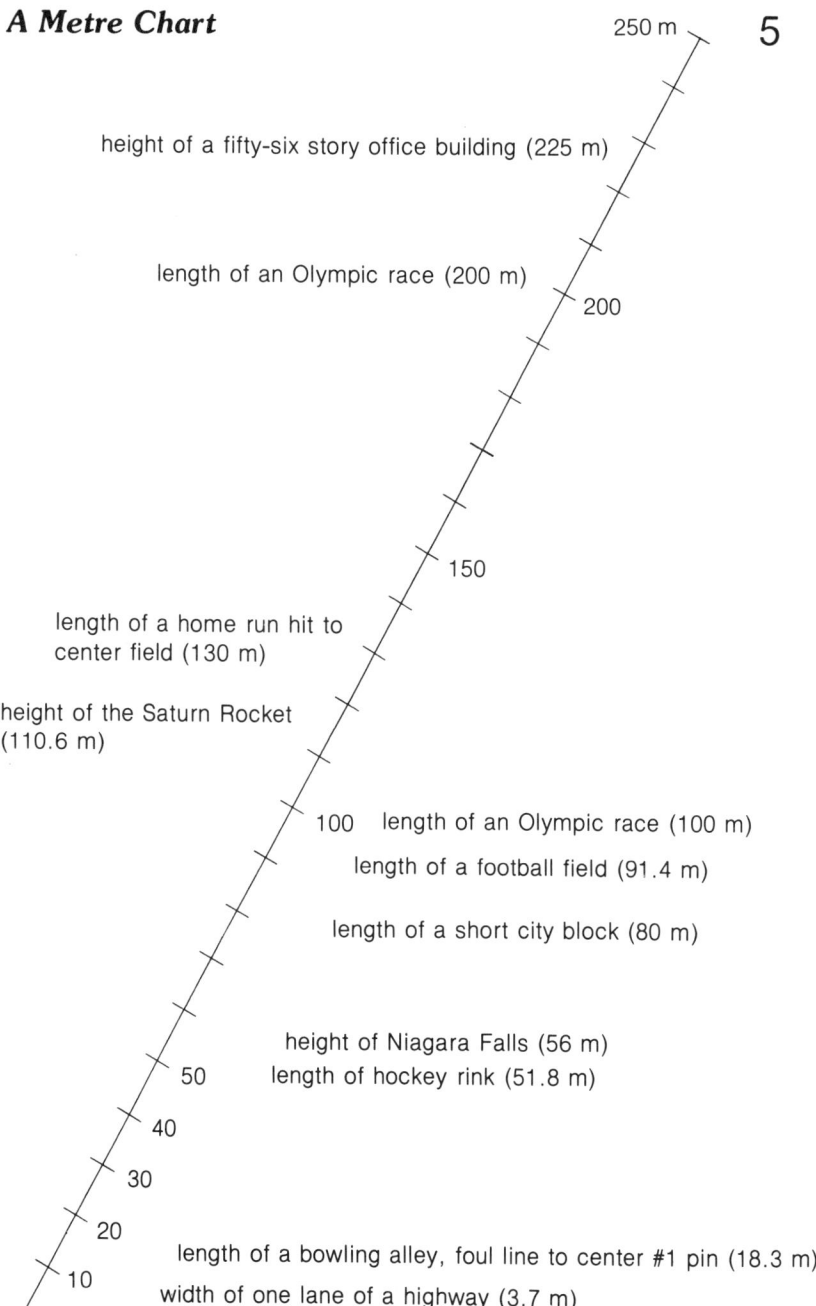

6 Practice the Metre

Here is a table to complete. Consider one item at a time. Estimate in terms of the metre and then check. Don't try to be efficient and do all the estimating and then all the checking. Do it our way and you will soon be estimating accurately.

Object	I think it is...			I was...	
	about the same length as a metre	longer than a metre	shorter than a metre	right	wrong
length of dresser					
height of stove					
width of door					
length of desk					
width of table					
length of rug					
width of rug					
width of window					

You should have no trouble with metres. Buying material or talking about the dimensions of a swimming pool or the distances in athletic events will be easy. The metre is under control.

The Centimetre

Examine your metre tape carefully. It has 100 equal parts. Each part is called a centimetre. The strip shown is twenty centimetres long and one centimetre wide. Five of these strips end to end equal one metre. The symbol cm is used for centimetres.

Remember
 1 metre = 100 centimetres

In symbols
 1 m = 100 cm

The centimetre is used for body sizes such as heights, waist and hip measurements, arm length and neck size. It is used for widths of textiles, dimensions of dress patterns and dimensions of furniture and applicances.

8 A Centimetre Chart

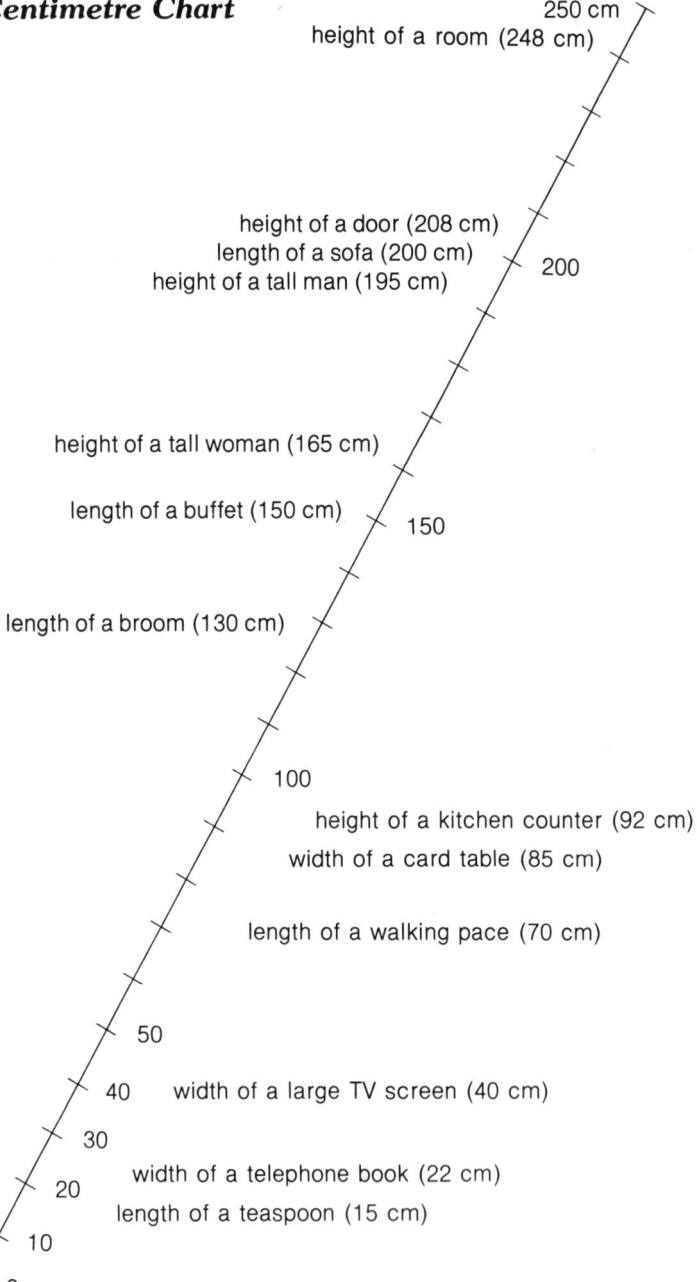

The metre tape is a bit long for measuring short objects. It would be handy to have a thirty-centimetre ruler. These are inexpensive, and it would help you to get a better feeling for the length of a centimetre. If you wish to make one, simply take a narrow strip of cardboard or wood and mark it off carefully in centimetres from zero to thirty, using your metre tape.

At this point you might like to compile some of your personal statistics. Cut a plain piece of cardboard 8 cm by 5 cm and draw up a chart like the one shown. This will fit into your wallet. When you complete the chart, you will be metricated.

```
            Me in Metric

   Name
   Height—           _____ cm
   Chest—            _____ cm
   Waist—            _____ cm
   Hips—             _____ cm
   Mass—          (to be added later)
```

10 Practice the Centimetre

Now let's think in terms of centimetres. Look at shorter objects than those you dealt with when you conquered the metre. Estimate how many centimetres long each object is. Then measure it with your metre tape or your centimetre ruler. Complete the table. Include other objects of your choice.

Object	I think it is ? cm long	It is ? cm long	The difference is ? cm
distance from elbow to tip of longest finger			
length of hand			
width of palm			
length of longest finger			
length of shortest finger			
width of finger nail			

The Kilometre

Another linear unit that you will need to know is the kilometre (km). The kilometre is used for measuring long distances. Since *kilo* means one thousand
$$1 \text{ kilometre} = 1000 \text{ metres}$$
In symbols
$$1 \text{ km} = 1000 \text{ m}$$
Road signs in most countries of the world give distances in kilometres. Very soon, signs in the United States will too. By that time, automobile manufacturers will be producing cars which will show speed in kilometres per hour (km/h). These cars will also use metric units for measuring engine capacity, oil capacity and gasoline capacity.

The concept of one kilometre is a little more difficult to grasp than that of one metre because it is a unit used for long distances. But since you have mastered the metre, you need only think in terms of 100 m to master the kilometre.

How long is a kilometre? To find out, measure 100 m of string with your metre tape. Persuade someone to hold one end, or tie it to a tree. Walk until the string is taut. Ten times this distance is a kilometre. The neighbors may wonder a bit at your actions, but if you do this you will have a good idea of one kilometre.

An average city block is about 100 m long. So if you walk about ten blocks, you will have walked one kilometre. Race courses for thoroughbreds range from 1.8 to 2.4 km. The city speed limit may be 56.5 km/h. A good par three golf hole might be about 165 m long, a par four 330 m and a par five perhaps 460 m. During a round of golf you might walk 6500 m or 6.5 km if you don't stray too much. On a bad day, add another kilometre.

Remember
$$1000 \text{ metres} = 1 \text{ kilometre}$$
$$1000 \text{ m} = 1 \text{ km}$$

The word "kilometre" is pronounced with the accent on the first syllable. This pronunciation is consistent with that of other metric units, such as the centimetre and the millilitre. Putting the accent on the second syllable is usually reserved for instruments of measurement, such as the speedometer and the thermometer.

12. Distances between cities will be measured in kilometres. Here is a chart of the approximate direct distances between familiar places.

Region	From	To	Distance in km
New York/ New England	New York City	Boston, MA	330
	Buffalo, NY	New York City	600
	Hartford, CT	New York City	175
	Portland, ME	New York City	495
	Hartford, CT	Boston, MA	160
Florida	Miami	Tampa	400
	Miami	Jacksonville	560
	Miami	Orlando	365
	Tampa	Jacksonville	310
	Jacksonville	Orlando	225
Chicago, IL	Chicago	Detroit, MI	430
	Chicago	Kansas City, MO	805
	Detroit, MI	Kansas City, MO	1200
Dallas/ Fort Worth, TX	Dallas/ Ft. Worth	Houston	390
	Dallas/ Ft. Worth	Amarillo	575
	Dallas/ Ft. Worth	El Paso	1000
	Dallas/ Ft. Worth	Austin	310
California	Los Angeles	San Francisco	610
	Los Angeles	San Diego	200
	San Francisco	San Diego	810
Across the United States	Los Angeles	New York City	4630

Practice the Kilometre   13

Now let's try some map reading. There are three maps for you to look at. The first map is of the Eastern United States, with the scale

1 centimetre = 50 kilometres

In symbols

1 cm = 50 km

To find the distance between any two places, measure the distance on the map in centimetres and change to kilometres. For example, the distance from New York, NY to Boston, MA, measures approximately 6.6 cm. Since

1 cm = 50 km

Then

6.6 cm = (50 X 6.6) km

So the distance is _____ km.

The second map is of the Western United States, with the scale

1 cm = 125 km

The third map is of the United States, with the scale

1 cm = 322 km

14 Use the maps to measure the distances between several cities and complete the table.

From	To	Distance in km

You can now smile when road signs show the distance in kilometres.

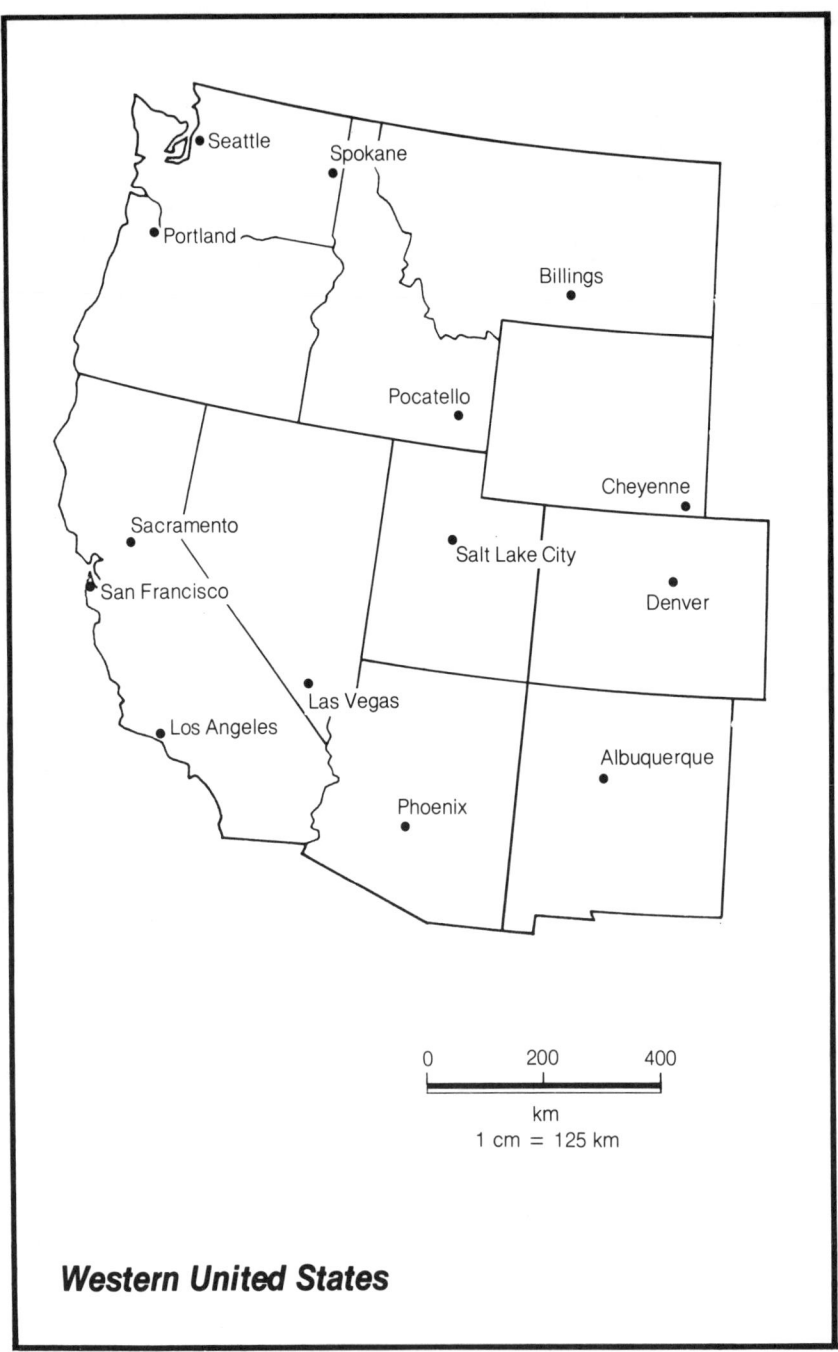

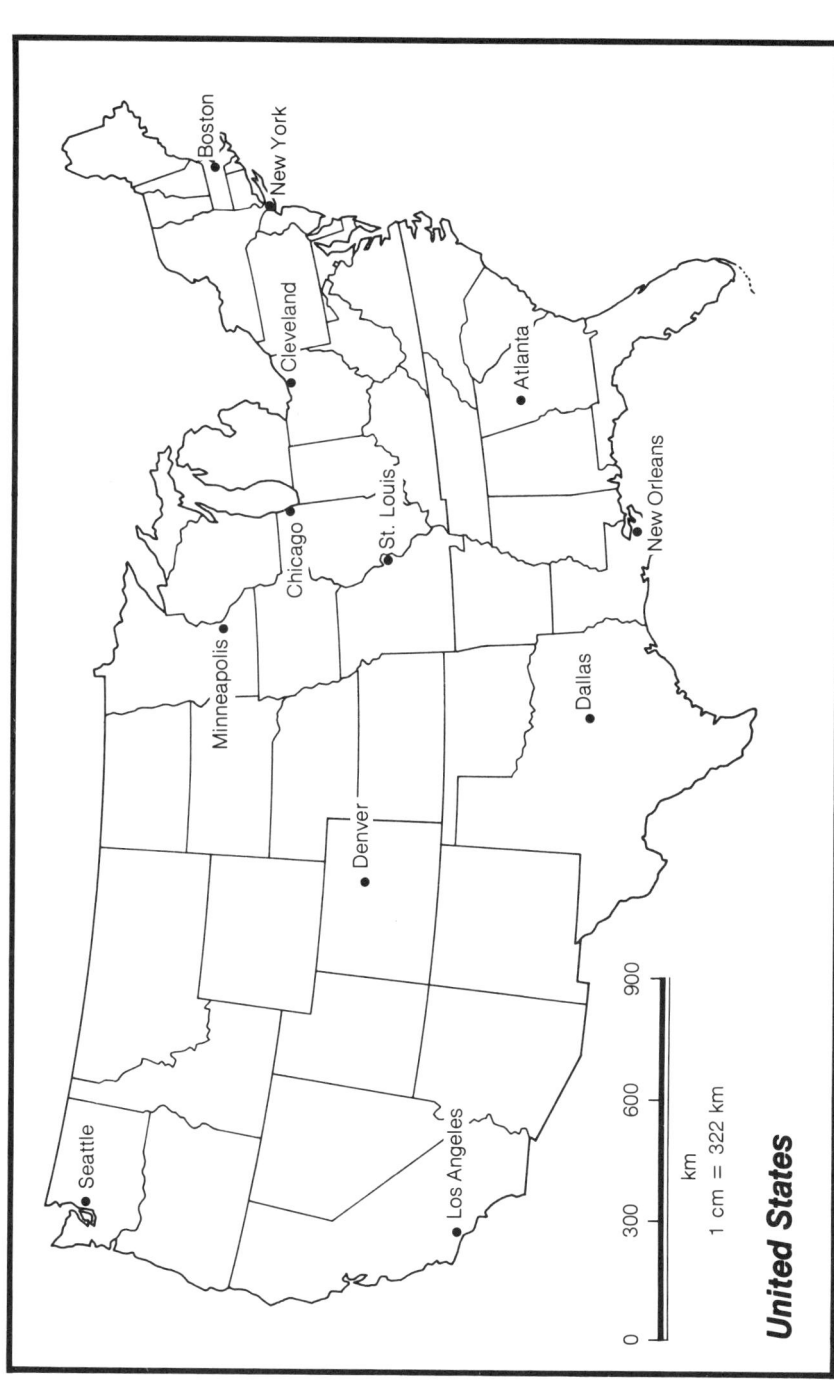

18 *The Millimetre*

The last unit of length you will need to know is the millimetre (mm). Since *milli* means one thousandth

1 millimetre = 1/1000 metre

In symbols

1 mm = 0.001 m

or

1 m = 1000 mm

If you examine a metric ruler used to make accurate measurements, you will notice that each centimetre is divided into ten equal parts. Each part is one millimetre.

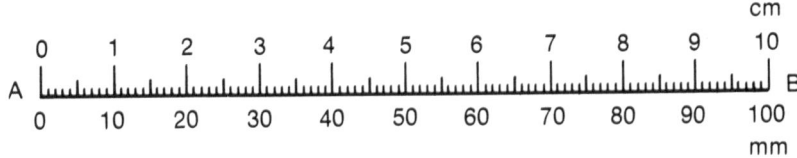

The line segment AB is 10 cm or 100 mm long. Each centimetre is divided into ten millimetres.

The millimetre is used to express dimensions of tacks, nails, hinges, screws, nuts and bolts; sizes of wrenches; thicknesses of glass, plywood, lenses, sheet metal; widths of ribbon, tape; amounts of rainfall; and so on. In scientific work, the millimetre is used exclusively, rather than the centimetre.

If you like to measure small things quite accurately, use a centimetre ruler graduated in millimetres.

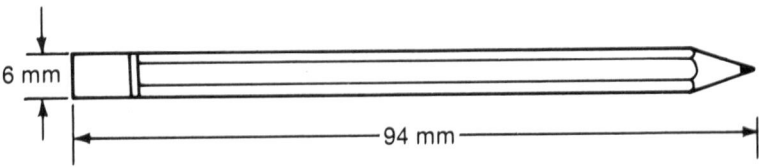

This pencil is 9.4 cm or 94 mm long. It is 0.6 cm or 6 mm wide.

Practice the Centimetre and the Millimetre   19

Measure each of the line segments and write the length in centimetres and also in millimetres. You can check your results with the answers at the back of the book.

Line segment	Length in cm	Length in mm
A—B C—D E—F G—H I—J P—Q R—S U—V X—Y Z—W	3.7 cm	37 mm

A ———————— B
    3.7 cm or 37 mm

(a) C ———————————— D

(b) E ———————————————— F

(c) G —— H

(d) I — J

(e) P ———————————————— Q

(f) R —————————————————— S

(g) U ———————————————— V

(h) X ——————————————————————— Y

(i) Z ————————————————————— W

20   When someone says a ribbon is 15 mm wide or gives another measurement in millimetres, you will now be able to visualize this width.

You have handled length very well. Just be sure to remember

$$10 \text{ mm} = 1 \text{ cm}$$
$$100 \text{ cm} = 1 \text{ m}$$
$$1000 \text{ m} = 1 \text{ km}$$

Can you decide how many millimetres are equal to a metre or to a kilometre? In a later section, we will see how to change from one unit to another. It's very easy to do — that is one of the advantages of the metric system.

Mass

Since metric units of mass (weight) will occur very frequently, we will discuss them next. You will need to become familiar with two units only, the kilogram and the gram. The kilogram is the basic unit of mass.

You may wonder why we have used the two words "mass" and "weight". Strictly speaking, there is a difference between these two words. An astronaut becomes weightless in space, but his mass does not change. While "mass" is the correct word for what we usually call "weight", the word "weight" will probably remain in common use.

The Kilogram

Since *kilo* means one thousand,
$$1 \text{ kilogram} = 1000 \text{ grams}$$
In symbols
$$1 \text{ kg} = 1000 \text{ g}$$

Let's get an idea of a kilogram first. You should buy a bathroom scale giving mass in kilograms only. If you see scales that show two systems of units, pass them by. Remember, you are thinking metric. Don't falter.

The kilogram is used in buying meat, vegetables and fruit. Marketing fruit by the dozen is somewhat unfair to the seller. The buyer naturally picks out the largest pieces, but pays no more than he would for the same number of smaller ones. When he buys by the kilogram, the buyer pays for the total mass of his selection, no matter what the size of the individual pieces. The kilogram is also used in buying goods such as grain, fertilizer and cement.

22 A Kilogram Chart

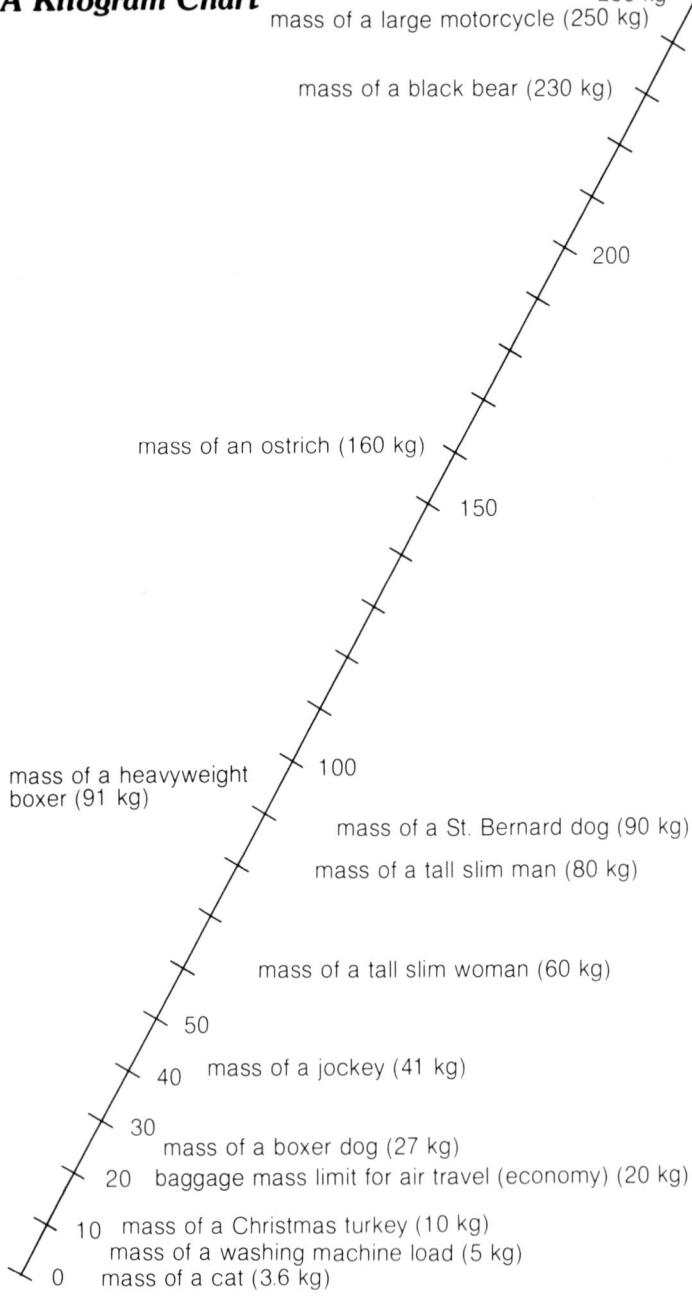

Practice the Kilogram   23

Record your mass in kilograms on your "Me in Metric" card. Metric bathroom scales will likely record mass to the nearest 0.5 kg (one-half). For example, a reading might be:

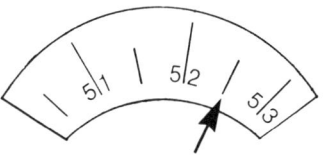

The mass is 52.5 kg (52 decimal 5 kg).

Estimate the mass of each member of your family in kilograms and then find the actual mass. Repeat this procedure (estimate and weigh) for any other people you can get to cooperate. Make a table and record your results in it. It is a good idea to record your own mass in a table each month, even though this can be disturbing.

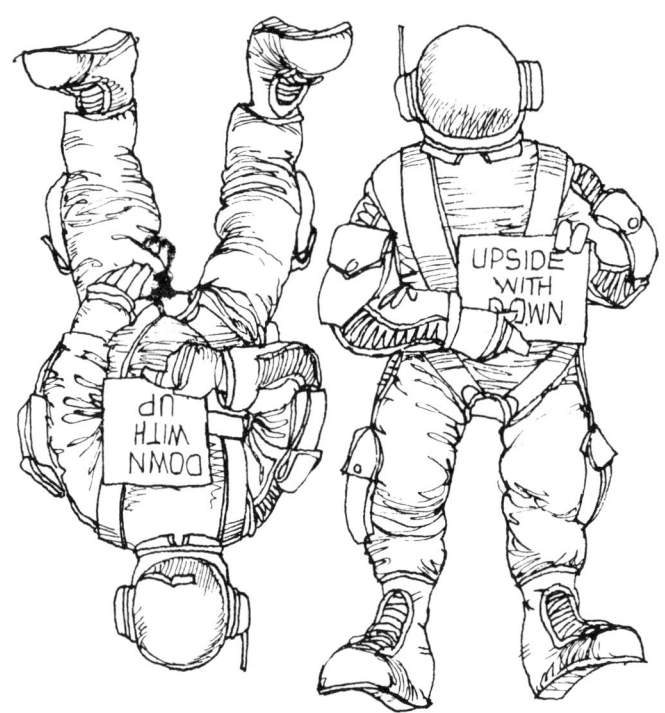

Name	Estimated mass in kg	Actual mass in kg

When you estimate the mass of a familiar object in kilograms, is your estimate fairly accurate?

Here is a height-mass chart for people with medium bone structure. Find your height in the left-hand column. Then find the corresponding mass in the appropriate column. Does your mass differ from the one given?

Height in cm	Mass of women in kg (approx.)	Mass of men in kg (approx.)	Mass of girls to age 18 in kg (approx.)	Mass of boys to age 18 in kg (approx.)
148	47.0		41.0	
150	48.0		43.0	
152	49.0	52.0	45.5	49.5
154	50.0	53.0	46.5	50.5
156	51.0	54.0	48.0	52.5
158	52.5	55.5	50.0	54.5
160	54.0	58.0	52.0	56.5
162	55.5	59.0	53.0	57.5
164	57.0	61.0	54.5	60.0
166	58.5	63.5	56.0	61.5
168	60.0	66.0	57.5	63.5
170	61.5	68.5	59.0	65.5
172	62.5	69.5	60.5	66.5
174	63.5	71.0	62.0	68.5
176	64.5	73.0	63.0	70.5
178	65.5	74.5	64.0	72.5
180	66.5	76.0	65.0	74.5
182	67.5	77.5	66.0	76.0
184		78.5		77.5
186		81.0		79.5
188		82.5		80.5
190		83.5		81.5
192		85.0		82.5
194		86.5		84.5
196		87.5		85.5

26 The Gram

A gram is a small unit of mass. Since the gram will be quite a common unit in the home, it would be handy to have a compression kitchen scale that records mass in grams. There are simple inexpensive models available, as well as more accurate types, and both can be used for a variety of purposes.

Although the best way to become familiar with the gram is to practice weighing objects, you can get along without a set of scales. Make a habit of estimating the masses of small products in the grocery or drug store and then check how close you are by looking at the labels. Here are some examples of the gram: three aspirins have a mass of about 1 g, a slice of bread has a mass of about 25 g, and a loaf of bread has a mass of about 700 g.

You might be wondering if there is a milligram unit of mass, since we use the millimetre unit for measuring length. The milligram is a very small unit of mass.

$$1000 \text{ milligrams} = 1 \text{ gram}$$

In symbols

$$1000 \text{ mg} = 1 \text{ g}$$

The milligram is used by the pharmacist for weighing the ingredients in pills.

HOW TO WEIGH A GRAM

A Gram Chart

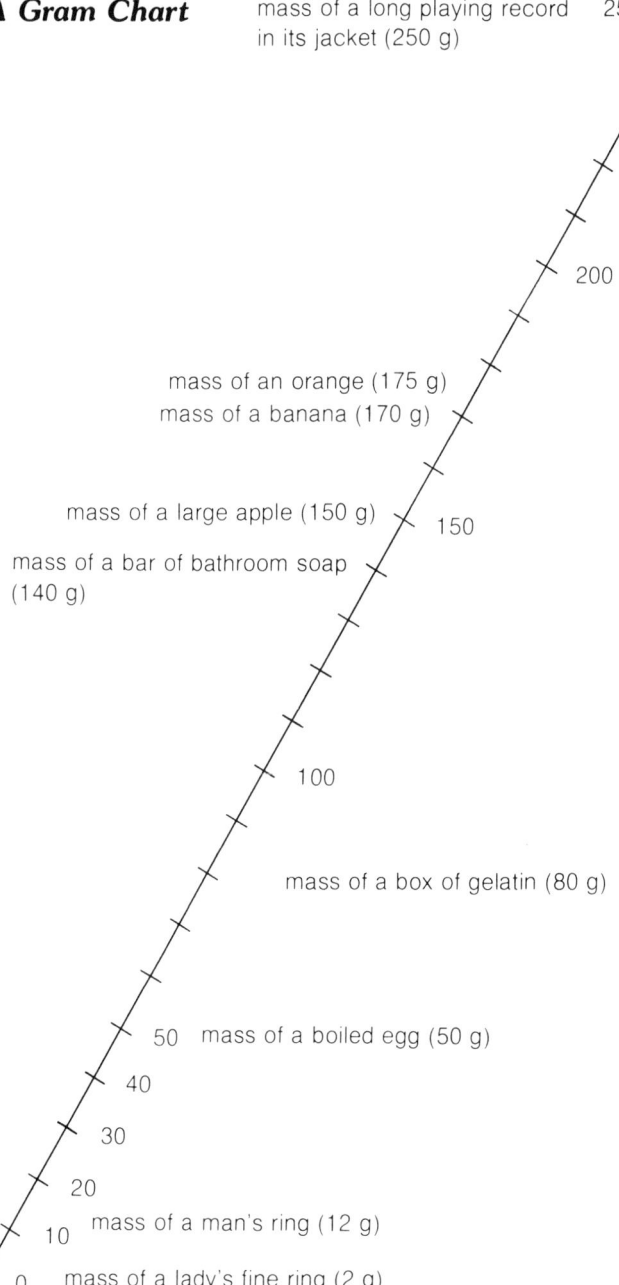

28 *Practice the Gram*

Use this table to keep track of your practice with the gram.

Date	Item	Cost per gram (check on prices)	Estimate in g	Actual mass in g

Capacity

Capacity tells us how much space there is inside a container. There are two units of capacity that you need to know, the litre (*l*) and the millilitre (ml). Since *milli* means one thousandth

$$1 \text{ millilitre} = 1/1000 \text{ litre}$$

In symbols

$$1 \text{ ml} = 0.001 \, l$$
$$\text{or}$$
$$1 \, l = 1000 \text{ ml}$$

Since there could be confusion between the sumbol and the numeral 1 (e.g., 121 l), the cursive *l* is often used to represent the litre.

The Litre

You will soon be buying milk by the litre. A large instant coffee jar holds about one litre of water. The litre is used to express the capacity of car radiators, gas tanks and household containers. Have you had a look at the litre chart?

30 The capacity of a container 10 cm by 10 cm by 10 cm is equivalent to 1ℓ.

$$\begin{aligned}\text{Capacity of container} &= (10 \times 10 \times 10) \text{ cm}^3 \\ &= 1000 \text{ cm}^3 \\ &= 1\ell\end{aligned}$$

Let's find the capacity of the bathtub in the litre chart. The dimensions of the tub are 150 cm by 65 cm and it is filled to a depth of 40 cm.

$$\begin{aligned}\text{Capacity of the tub} &= (150 \times 65 \times 40) \text{ cm}^3 \\ &= 390,000 \text{ cm}^3\end{aligned}$$

Since we know that 1000 cm³ = 1ℓ, the capacity of the tub in litres

$$= \frac{390,000}{1000} \ell$$

$$= 390 \ \ell$$

You can buy a litre container at very little cost.

A Litre Chart

capacity of a tank 100 cm by 100 cm by 50 cm (500 ℓ) — 500 ℓ

capacity of a bathtub 150 cm by 65 cm filled to a depth of 40 cm (390 ℓ) — 400

capacity of a big refrigerator (300 ℓ)

capacity of a vat 60 cm in diameter by 100 cm high (270 ℓ) — 300

200

100

80

60 — capacity of an aquarium 70 cm by 35 cm by 25 cm (61.25 ℓ)

40

20 — capacity of a crock 30 cm in diameter by 50 cm high (34 ℓ)

capacity of a pail (10 ℓ)

0 — capacity of a tea kettle (2 ℓ)

31

32 The Millilitre

The millilitre (ml) is a small unit of capacity. A small thimble holds about 2 ml of water.

One millilitre of water has a mass of one gram. The scientist would also specify the temperature of the water but we won't worry about it. This relationship between the capacity and mass of water gives us a different way of measuring capacity. You may want to explore this idea further. Weigh a container filled with water. Let's say you find that it weighs 48 kg. Then weigh the empty container. You may find that it weighs 3 kg. The mass of the water is therefore 45 kg. The capacity of the container is 45 ℓ since 1 ml of water weighs 1 g, 1 ℓ of water weighs 1 kg.

It would be handy to have a container marked in millilitres. Select a uniformly shaped container such as a plastic or glass bottle. (Put tape on the glass bottle to facilitate marking.) A container 5 cm square by 40 cm high could be marked off in 4-cm increments to show each 100 ml.

$$5 \text{ cm} \times 5 \text{ cm} \times 4 \text{ cm} = 100 \text{ cm}^3 = 100 \text{ ml}$$

The 4-cm increments can then be marked with 10 even divisions to delineate 10 ml increments.

$$5 \text{ cm} \times 5 \text{ cm} \times \frac{4 \text{ cm}}{10 \text{ divisions}} = 10 \text{ cm}^3 = 10 \text{ ml}$$

Your 5 cm square container need not be 40 cm tall—only tall enough to mark several 4-cm increments to provide a convenient measuring vessel.

In general, the millilitre is used to express the capacity of containers holding less than a litre. Volumes of medicines, shampoos, lotions and other liquids sold in relatively small quantities can be conveniently expressed in millilitres.

While you are grocery shopping, see how many different containers you can find which are marked in millilitres. Estimate the capacity of each before you check the label.

A Millilitre Chart

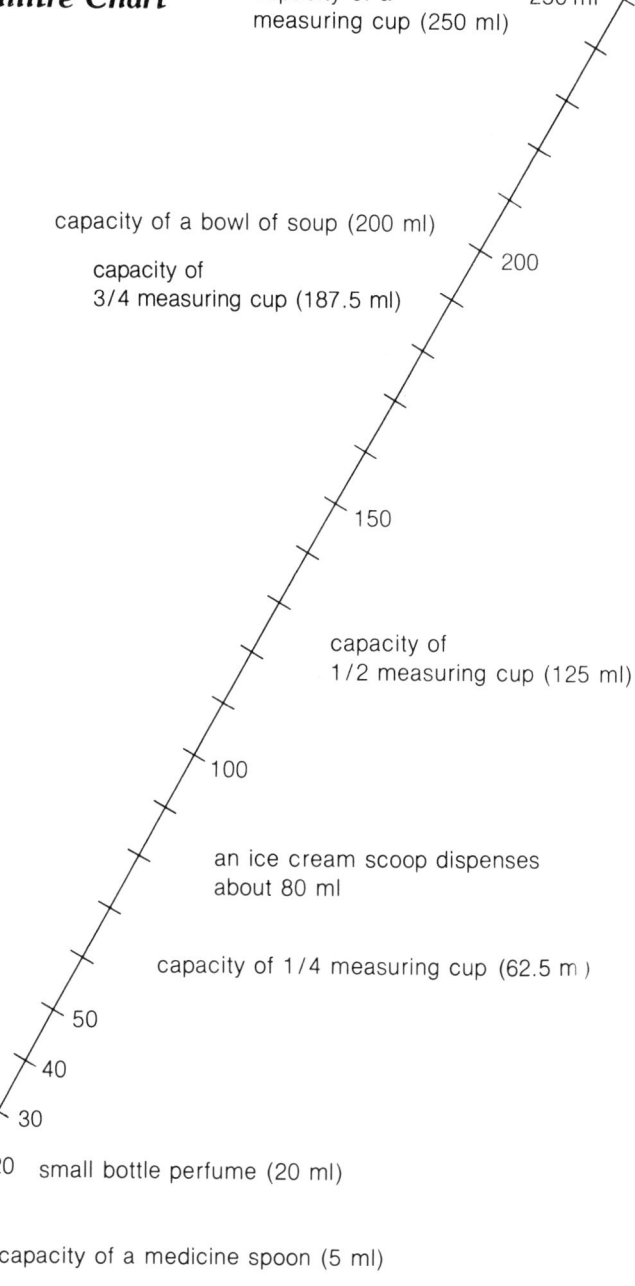

33

34 Practice the Litre and the Millilitre

It can be fun to estimate the capacities of various containers — tall thin ones, short stout ones, odd-shaped ones.

Try to estimate capacities to the nearest 0.5 ℓ (one-half), using many containers of different shapes: mixing bowls, tea pots, vases, and pails. Use your container marked in millilitres to check your estimate of the smaller containers.

Estimate, measure and record your results in the table. The more you do, the better will be your ability to estimate which product is a good buy when you are comparison shopping.

Container	Estimated capacity in ml	Actual capacity in ml	Actual capacity in ℓ
A			
B			
C			
D			
E			
F			
G			
H			

Area 35

Area is a two-dimensional measurement involving length and width. You are familiar with tiles covering a floor. We could say that the area of the floor is so many tiles, but this wouldn't be a standard unit of measurement.

The Square Centimetre

We measure length in centimetres (cm) or metres (m). We measure area in square centimetres (cm^2) or square metres (m^2). In the symbol cm^2, the raised 2 tells us the number of dimensions. Here is a square centimetre. It is 1 cm long and 1 cm wide.

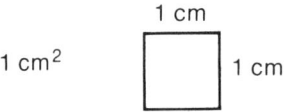

The square centimetre is a small unit of area. It is used to describe the area of cards, photographs, tiles, newspaper advertisements and maps.

36 A Square Centimetre Chart

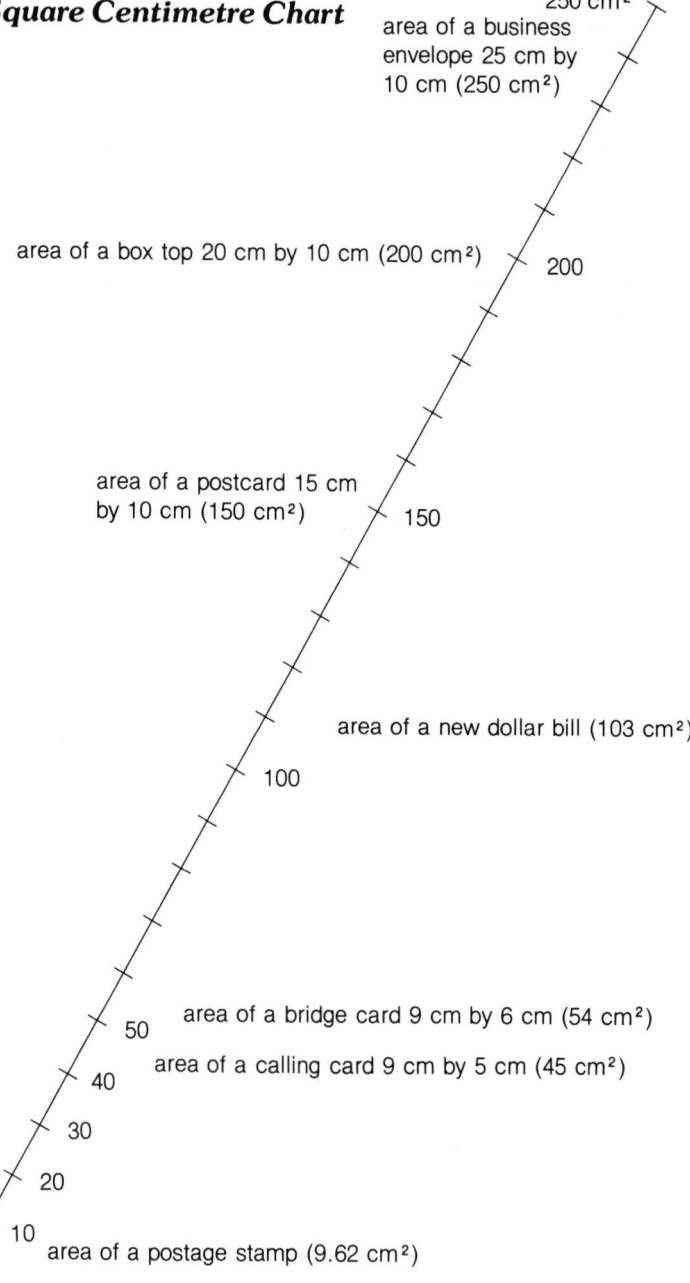

A good way to understand square units is to look at areas enclosed by squares and rectangles. This practice will also stand you in good stead if someone asks you for help with mathematics homework!

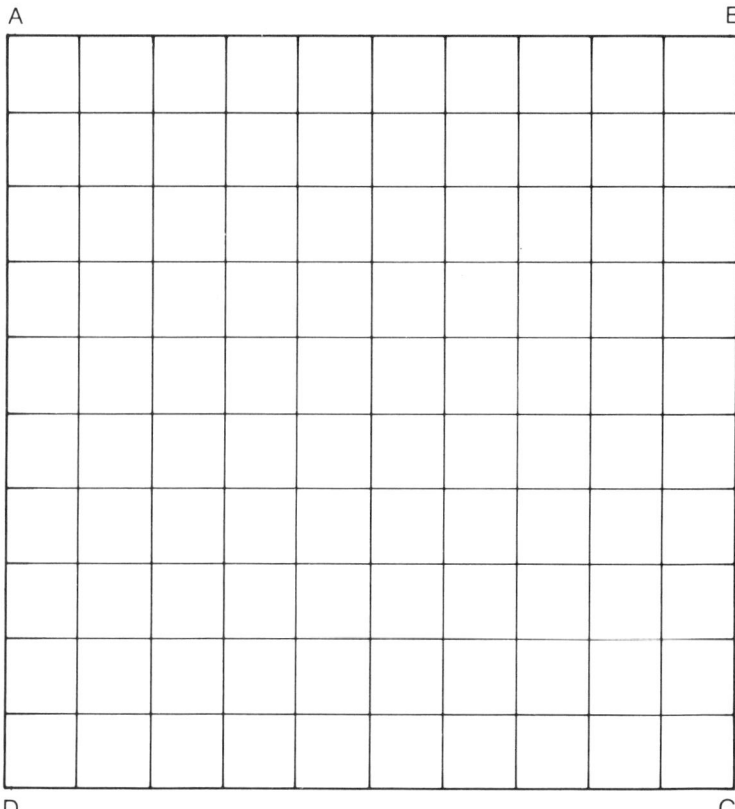

38 The square ABCD is made up of four equal line segments AB, BC, CD and DA, and four square corners. We find the area enclosed by the square this way:

$$\text{Area of square ABCD} = \text{length} \times \text{width}$$
$$= (10 \times 10) \text{ cm}^2$$
$$= 100 \text{ cm}^2$$

To check this answer, count the small squares. Each is 1 cm². We find that there are 100 small squares or 100 cm².

Now let's look at a rectangle. This time we have four square corners, but all sides are not equal.

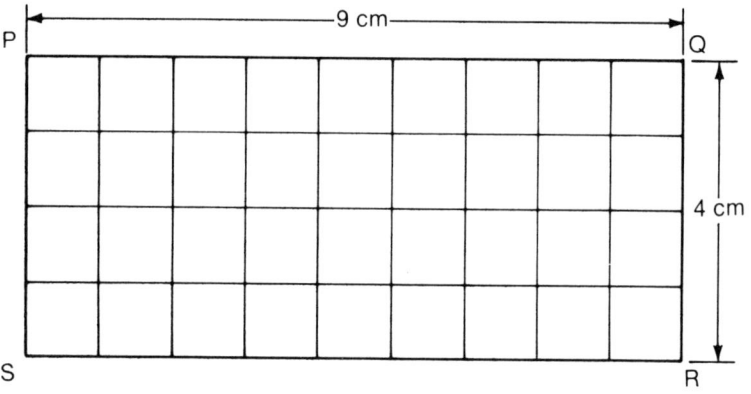

$$\text{Area of rectangle PQRS} = \text{length} \times \text{width}$$
$$= (9 \times 4) \text{ cm}^2$$
$$= 36 \text{ cm}^2$$

Practice the Square Centimetre

Using your centimetre ruler, mark each of the following figures in square centimetres. Record the area as we have done in the examples you have just looked at. You may check by counting squares or by comparing your answers with those at the back of the book.

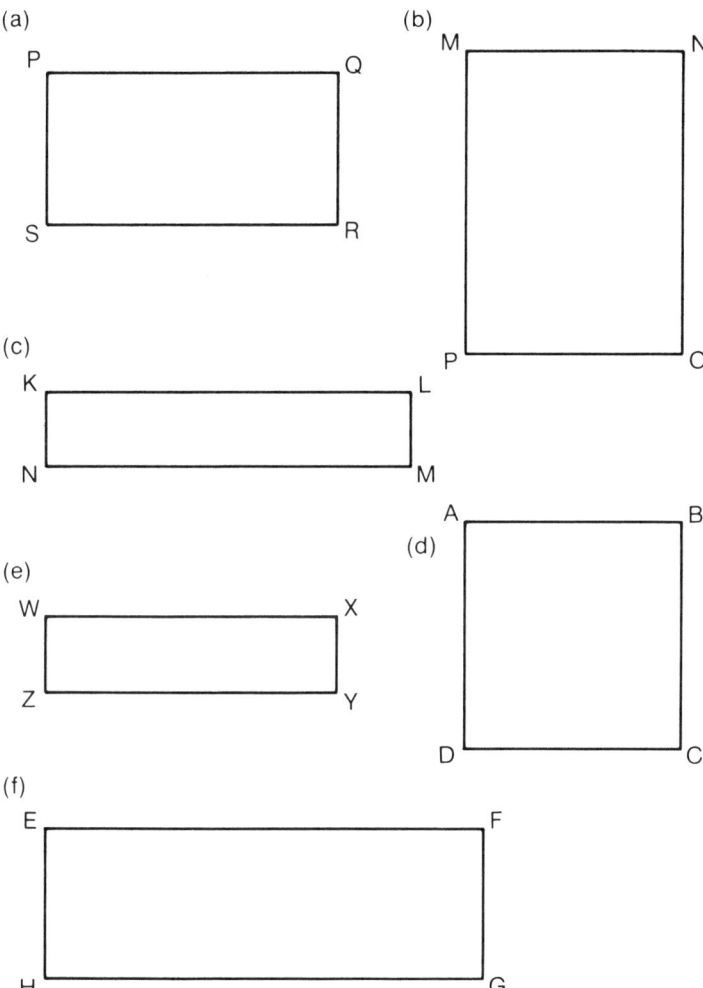

40   The Square Metre

The symbol for the square metre is m². The square metre is a large unit of area compared to the square centimetre. It would be well worth the trouble to take a look at a square metre. Draw one on the sidewalk, basement floor or driveway by using a straight edge, your metre tape and a piece of chalk. Use a sheet of cardboard or a book to get the corners square. Your square metre will be 1 m or 100 cm long and 1 m or 100 cm wide.

$$1 \text{ m}^2 = (100 \times 100) \text{ cm}^2$$
$$= 10{,}000 \text{ cm}^2$$

Areas of walls, floors, patios and driveways are given in terms of square metres.

Can you estimate the area of the wall next to you? If you think it is 5 m long and 3 m high, then its

$$\text{Area} = \text{length} \times \text{width}$$
$$= (5 \times 3) \text{ m}^2$$
$$= 15 \text{ m}^2$$

When you check, you will probably find that the dimensions are not an exact number of metres. You might measure each dimension to one decimal place. Have you forgotten how to handle decimals? We have done an example of multiplication with decimals following the square metre table. If the procedure still isn't clear, check the review of decimals at the end of the book.

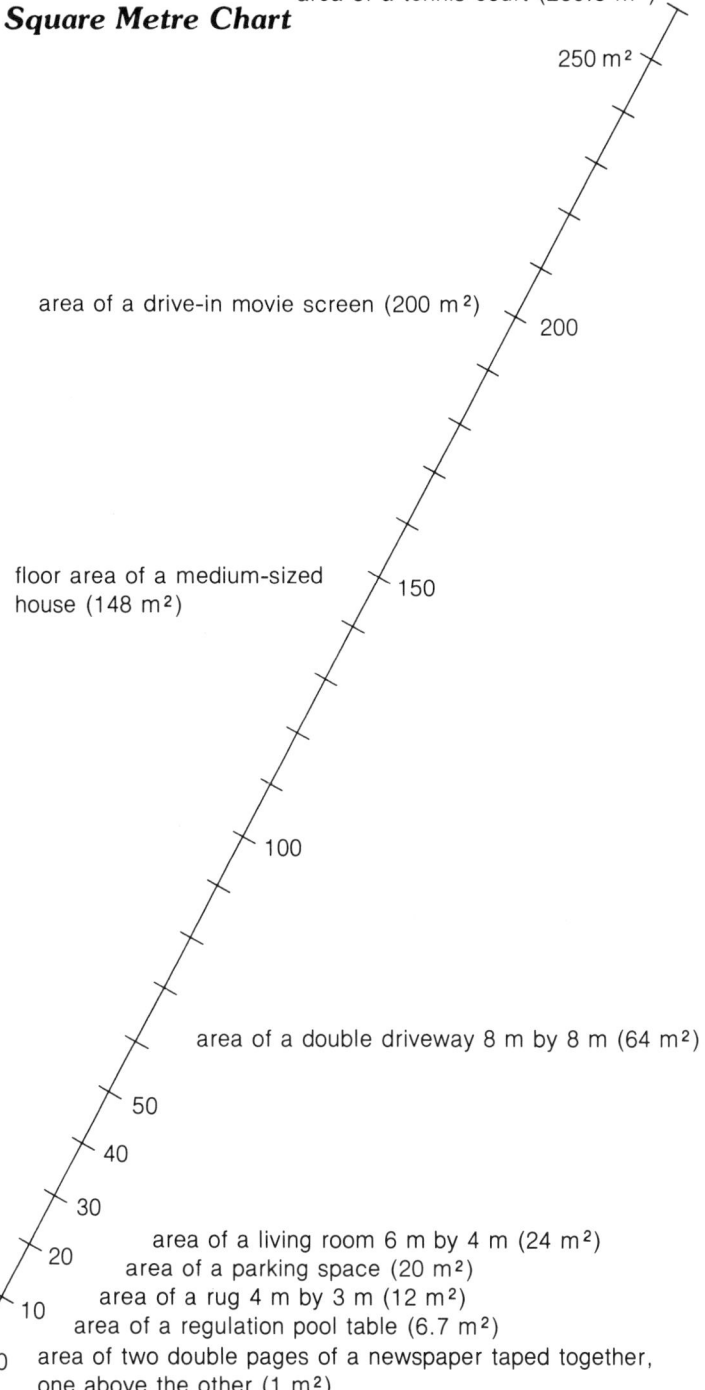

42 Practice the Square Metre

This table requires quite a bit of estimating, measuring and multiplying in decimals. The first row has been completed as an example.

Object	Estimated			Measured		
	length in m	width in m	area in m²	length in m	width in m	area in m²
table	2	1	2	1.8	0.9	1.62
floor						
wall						
rug						
top of stove						
blanket						
bed						
picture						
door						
end table						

```
  1.8   ← one place after the decimal point
X 0.9   ← one place after the decimal point
  ‾‾‾
  1.62  ← two places after the decimal point in the product
```

The Hectare

Have you wondered what unit will be used for measuring large land areas, such as fields and farms? The hectare (ha) will be used for this purpose. The hectare is an area 100 m by 100 m or 10,000 m².

We certainly don't expect you to do a lot of surveying to get an idea of the size of a hectare; however, a little thought and some experimenting will give you a pretty good idea. Check the length of your normal walking pace, using the metre tape. For example, you might find that you take three steps for two metres. Then count the number of steps you take in walking an ordinary city block. When you have walked 150 steps, you have gone 100 m.

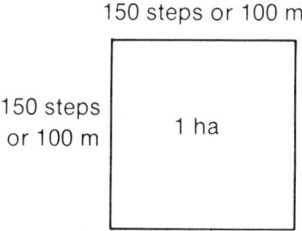

If the block is about 150 steps by 150 steps, its area is 1 ha. You might prefer to try measuring out 150 steps in a large field or park. From time to time, you will hear that some familiar region has an area of so many hectares, for example, Washington, DC, has an area of about 17,353 ha, and the Grand Canyon, AZ, has an area of about 272,596 ha.

44 A Hectare Chart

area of four average city blocks 160 m by 160 m (2.5 ha)

area of a parking lot 200 m by 100 m (2 ha)

area of ten building lots, each 60 m by 30 m (1.8 ha)

area of a field 100 m by 100 m (1 ha)

area of a football field (0.45 ha)

area of a house lot 30 m by 18 m (0.05 ha)

Volume

Volume is a three-dimensional measurement involving length, width and height. Like capacity, volume tells how much space an object occupies. It is easy to decide whether to use units of capacity or units of volume for measuring. Common containers such as those used for relatively small quantities of liquids are measured in units of capacity such as litres and millilitres; whereas, other quantities are measured in volume.

The Cubic Centimetre

We use the centimetre (cm) to measure length, the square centimetre (cm^2) to measure area, and the cubic centimetre (cm^3) to measure volume. The cubic centimetre is a small unit 1 cm long, 1 cm wide and 1 cm high. It is used to describe the volumes of small boxes. It is often used in mathematics to describe the volumes of cones, cylinders and spheres.

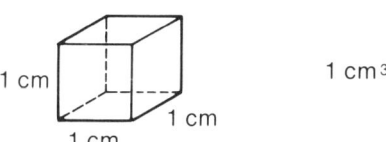

Measure the length, width and depth of a rectangular box and find the volume in cubic centimetres. The one we tried was 23 cm by 20 cm by 10 cm.

$$\begin{aligned} \text{Volume} &= \text{length} \times \text{width} \times \text{depth} \\ &= (23 \times 20 \times 10) \text{ cm}^3 \\ &= 4600 \text{ cm}^3 \end{aligned}$$

The volume of an attaché case 43 cm by 30 cm by 6 cm is 7740 cm^3.

46 A Cubic Centimetre Chart

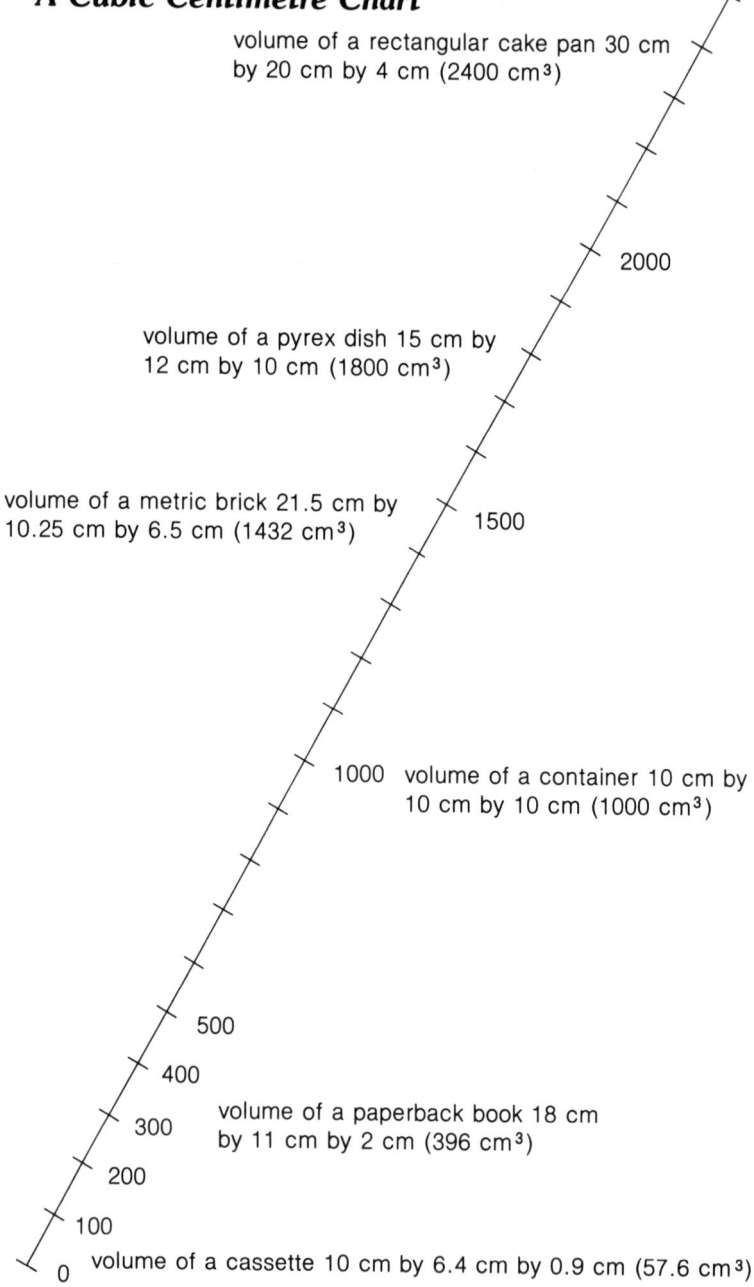

Practice the Cubic Centimetre 47

If you measure a number of different rectangular boxes and pans, you will quickly get an idea of volume in terms of cubic centimetres. Record your results in the table.

Object	Estimated volume in cm³	Calculated volume in cm³
small rectangular box a bigger rectangular box bread box FM speaker enclosure FM stereo component small dresser drawer		

48   Calculate the volume of these figures in cubic centimetres. Record your calculations as in the example we looked at earlier. You can check your answers with those at the back of the book.

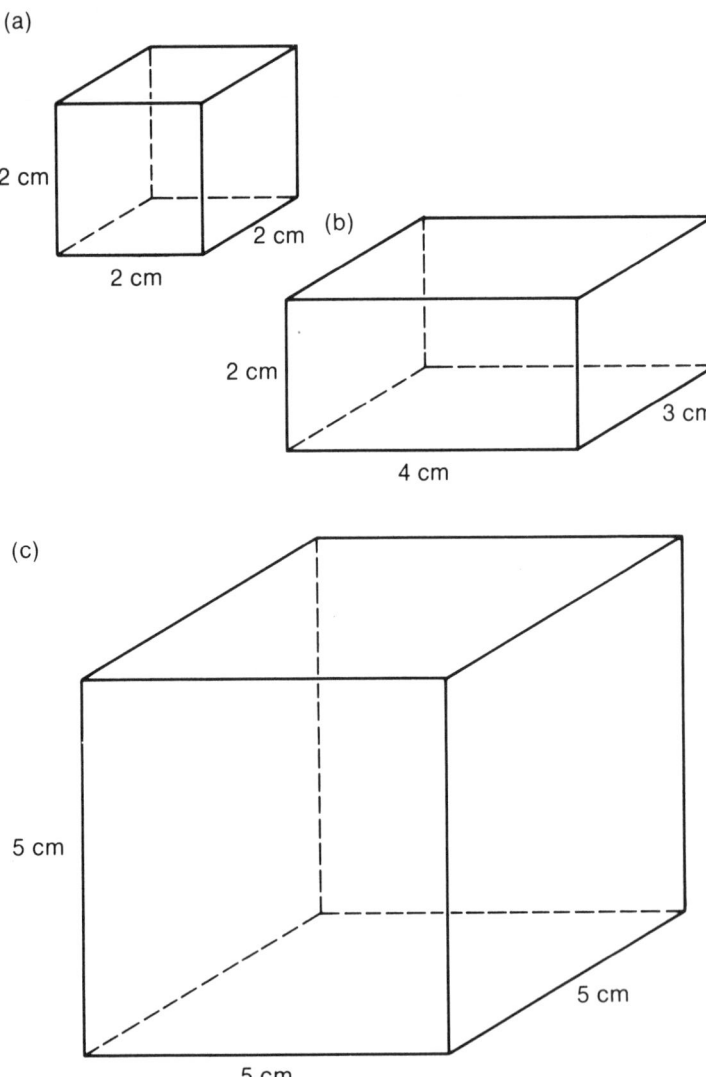

The Cubic Metre

We use the metre (m) to measure length, the square metre (m²) to measure area, and the cubic metre (m³) to measure volume. The cubic metre is a large unit of volume. It is 1 m or 100 cm long, 1 m or 100 cm wide and 1 m or 100 cm high.

$$1 m^3 = (100 \times 100 \times 100) \ cm^3$$
$$= 1,000,000 \ cm^3$$

Gravel and mixed cement are sold by the cubic metre. The volume (or capacity, depending on the cargo) of a dump truck is given in cubic metres.

It is worth the trouble to build a model of a cubic metre. You can do this with twelve thin wooden strips, each 1 m long, and eight wooden blocks, with holes carefully drilled, to form the corners.

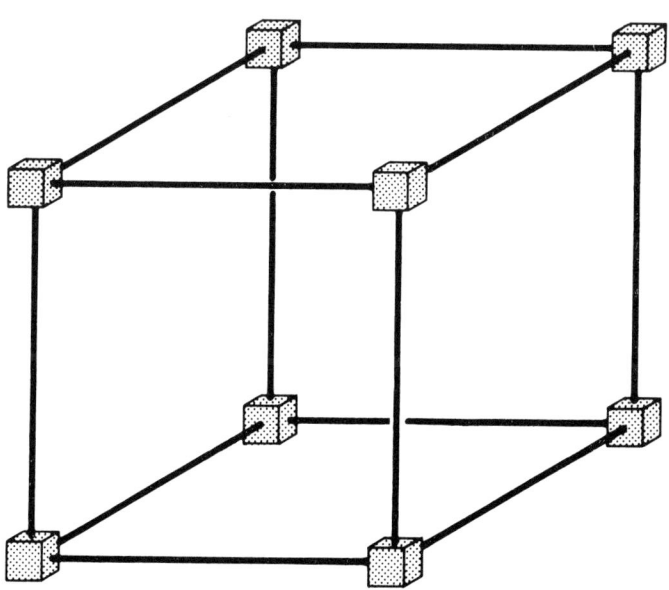

50 A Cubic Metre Chart

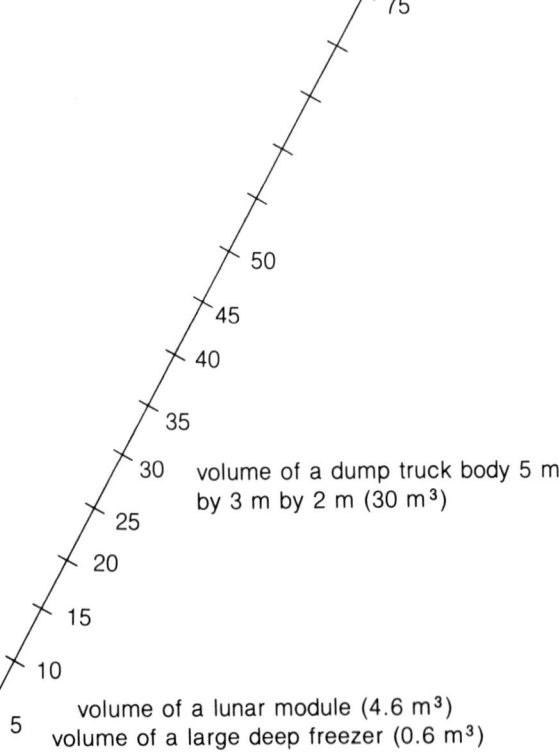

125 m³

volume of a ditch 12 m by 5 m by 2 m (120 m³)

volume of a family swimming pool 12 m by 6 m with an average depth of 1.5 m (108 m³)

100

volume of a living room 7 m by 5 m by 2.5 m (87.5 m³)

75

50
45
40
35
30   volume of a dump truck body 5 m by 3 m by 2 m (30 m³)
25
20
15
10
5   volume of a lunar module (4.6 m³)
     volume of a large deep freezer (0.6 m³)
0   volume of a refrigerator (0.35 m³)

Practice the Cubic Metre

Here is a table to complete. One example is done for you.

Object	Length in m	Width in m	Height in m	Volume in m³
a bin	2.3	1.5	0.8	2.3 X 1.5 X 0.8 = 2.76
your room				
a dump truck body				
a deep freezer				
a large drawer				
a closet				
an excavation for a house 15 m by 9 m by 2 m				

The calculation for the bin:

```
  2.3          3.45
 X1.5         X 0.8
 ---          -----
 115          2.760
  23
 ----
 3.45
```

52 *Temperature*

The Celsius Thermometer

Becoming familiar with the Celsius thermometer and metric temperatures will not be a problem. Learn the three important temperatures listed on the right-hand side of the temperature scale illustrated. Listen to temperatures (highs and lows) given in Celsius on the weather reports. But most important, purchase a Celsius thermometer. Don't let anyone sell you a thermometer with two scales. Anyone who tries to is probably concerned about getting rid of them before the Celsius scale is the only one in demand!

    The symbol for temperature in Celsius is °C. Here we do use a capital letter because the C stands for a man's name; the unit is the degree. So this is really not an exception to the rule of lower case letters for unit symbols.

Celsius Temperature Scale

53

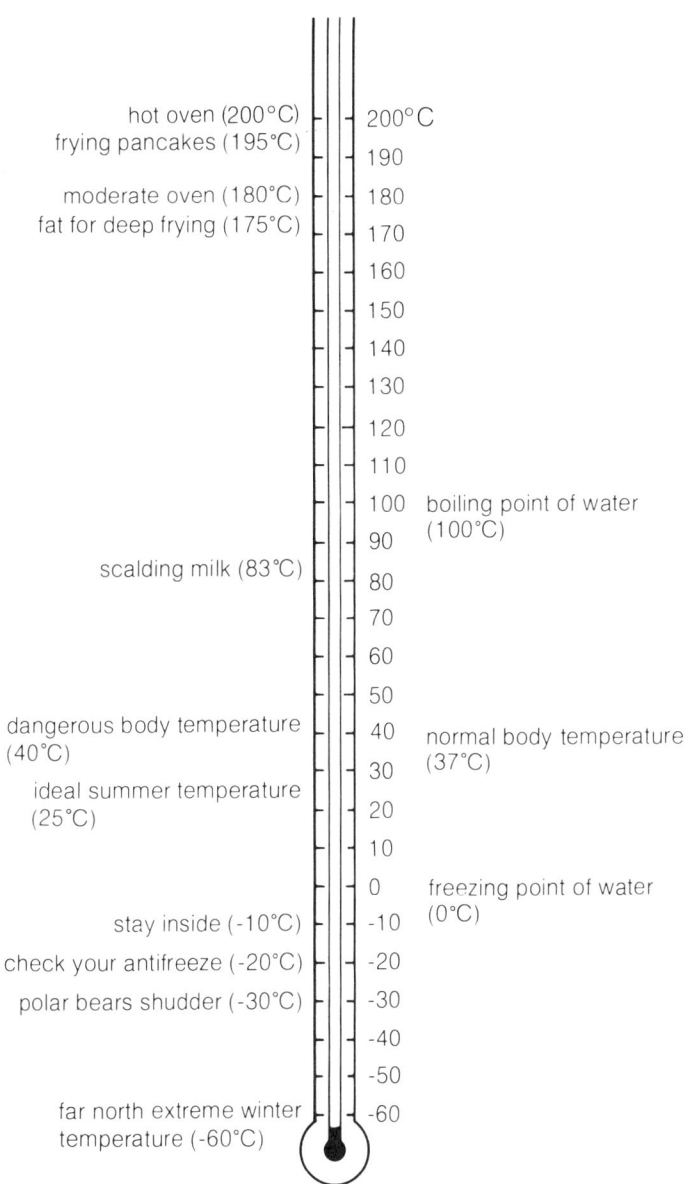

54 The Clinical Thermometer

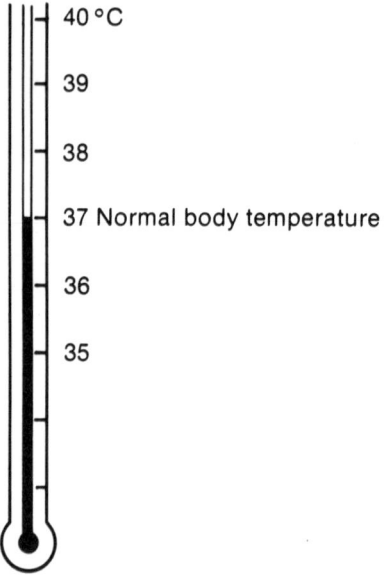

You should have a clinical Celsius thermometer. Don't go to work if your temperature goes up past 37°C. Normal body temperatures for animals are different from ours. Your dog's normal temperature is 38.8°C. But for you:

        36°C is subnormal
        37°C is normal
        38°C is feverish
        39°C is very feverish
        40°C is dangerous

Consumer Goods 55

You should now be able to picture the following metric units.

Length	Area	Volume	Mass	Capacity
mm cm m km	cm² m² ha	cm³ m³	g kg	ml *l*

Since you now understand the correct use of these units, you will be able to compare costs for different kinds of packages when you are shopping. No one will have to convert for you. You know the metric system of measurement. Here are some examples of common goods and the units likely to be used in connection with them.

Category	Item	Unit
clothing	ready-made clothing for women	not affected – sizes 12, 14, 16, etc., as usual
footwear	shoes, etc.	sizes as usual – lengths and widths in cm
furnishings	drapes carpets	length in m area in m²
groceries pre-packaged groceries	meats, fruits, vegetables bread milk butter canned goods sugar, tea, coffee	mass in kg mass in g capacity in *l* mass in g capacity in ml mass in g

hardware	nails, screws, bolts paint	length in mm capacity in ℓ
lumber	paneling uprights flooring	length in m width in m length in m width in cm length in m width in cm
drugs	cough syrup vitamin pills	capacity in ml mass in mg
photographic equipment	film	width in mm
postage rates		based on g and kg – the maximum mass for international parcels is 10 kg

Recall of Decimals

Quiz

Try this brief quiz. Check your answers with those at the back of the book. If you have difficulty with a certain type of question, you will know that you should study that section of the review on the next few pages.

1. In 5037.6
   What does the 3 tell you?
   What does the 0 tell you?

2. Complete
   (a) 49.3 X 10
   (b) 57.6 ÷ 10
   (c) 0.006 X 1000
   (d) 0.006 ÷ 10
   (e) 576.4 ÷ 100
   (f) 1.11 X 100

3. Add   86.1
           234.35
             9.85
            12.06

4. Subtract 576.04
               87.35

5. Find the total amount $ 53.29
                                843.25
                                  9.68
                            1476.52

6. Find the difference in amount $9356.04
                                            986.27

7. Multiply 43.5
          0.6

8. Multiply 207.32
           4.9

9. Divide 38)129.2

10. Divide 53.7)461.820

Operations with Decimals

Add 69.2, 138.29, 6.05, 7105.698, 399.001. Simply line up the decimal points in a column and add.

```
  69.2
 138.29
   6.05
7105.698
 399.001
7718.239
```

Subtract 935.889 from 6504.793. Again, just keep the decimal points in line and subtract.

```
6504.793
 935.889
5568.904
```

Multiply 4.8 X 5.

4.8 ← *one place after the decimal point*
  5
24.0 ← *one place after the decimal point*

Multiply 23.67 X 4.3 (there are two factors: 23.67 and 4.3).

23.67 ← *two places after the decimal point*
 4.3 ← *one place after the decimal point*
7101
9468
101.781 ← *three places after the decimal point*

Count the number of places after the decimal point in each factor. The sum of these two numbers is the number of places after the decimal point in the answer.

Divide 37.68 by 23.

```
         1.63
    23) 37.68
        23
        146
        138
         88
         69
         19
```

23) 37̇.68 ← the decimal in the answer is lined up over the decimal in the question

no decimal here

60   Divide 589.65 by 4.7.

```
                                            125.4
4.7)589.65        47)4896.5         47)5896.5
  ↑                 ↶   ↶              47
 we don't want a   move both           ---
 decimal here      decimals one        119
                   place to the         94
                   right               ---
                                       256
                                       235
                                       ---
                                       215
                                       188
                                       ---
                                        27
```

Divide 723 by 1.23.

```
                                            587.80
1.23)723          123)72300.        123)72300.00
   ↑                 ↶    ↶            615
 two places of    move both            ----
 decimal here     decimals two         1080
                  places to the         984
                  right                ----
                                        960
                                        861
                                        ---
                                        990
                                        984
                                        ---
                                         60
```

You may wonder why we can move the decimal point in this way. We can write any division question as a fraction.

589.65 ÷ 4.7   can be written $\frac{589.65}{4.7}$

We can multiply the top and bottom of a fraction by any number, as long as we multiply *both* top and bottom by the *same* number. We could use 10, 100 or 1000, but 10 will do.

$$\frac{589.65}{4.7} \times \frac{10}{10} = \frac{5896.5}{47}$$

The result is equivalent to moving both decimals one place to the right.

Multiplying and Dividing by 10, 100 and 1000

To multiply or divide by 10, 100, 1000, etc., we simply shift the decimal point as many places as there are zeros.

53.8 X 10 = 538
shift one place to
the right

53.8 ÷ 10 = 5.38
shift one place to the
left

.0074 X 100 = 0.74
shift two places to
the right

.0074 ÷ 100 = .000074
shift two places to the
left

570.005 X 1000 = 570,005
shift three places to the
right

570.005 ÷ 1000 = .570005
shift three places to the
left

Since the metric system is a decimal system, a measurement expressed in one unit can be expressed in a smaller unit by multiplying by 10, 100, 1000, etc. A measurement is expressed in a larger unit by dividing by 10, 100, 1000, etc. This is done by shifting the decimal point as indicated.

Complete the following table. Think: if you are changing to a smaller unit there should be more of them — better multiply! Check your answers with those at the end of the book.

Measurement	Change to	Relationship	Result
(a) 50 cm	m	100 cm = 1 m (÷100)	0.5 m
(b) 43.6 km	m	1000 m = 1 km (×1000)	43,600 m
(c) 67.8 cm	m	100 cm = 1 m	
(d) 42.6 m	km	1000 m = 1 km	
(e) 43,658 cm	km		
(f) 756 ml	ℓ		
(g) 3561 ℓ	ml		

Practice Exercise

Check your answers with those at the back of the book. Has your performance improved since the quiz?

1. Add
(a) 0.41 (b) 5.23 (c) 4.01 (d) 14.00 (e) 47.62
     2.00   17.00   2.00   4.32   8.93
     0.06   4.00   4.10   1.62   124.67
     1.98   0.27   41.00   46.30   39.84

2. Subtract
(a) 4.347 (b) 14.000 (c) 5.300 (d) 231.60 (e) 5476.235
     1.259   2.536   1.006   117.43   3837.146

3. Multiply
(a) 13.5 (b) 467.9 (c) 5.875 (d) 6.857
      1.1   0.8   3.25   7.394

4. Divide

(a) 5.63) 15.201   (b) 23.9) 1142.42   (c) 2.05) 63.632

Charts and Graphs

Here is a mass-length chart for female babies from birth to one year. Masses are given in kilograms and lengths are given in centimetres.

	Girls				
	very low	low	average	high	very high
Birth					
Mass	under 1.7 kg	1.7 - 3.2	3.2 - 3.6	3.6 - 4.3	over 4.3
Length	under 47 cm	47 - 49	49 - 51	51 - 53	over 53
3 months					
Mass	under 4.4 kg	4.4 - 5.3	5.3 - 6.0	6.0 - 6.8	over 6.8
Length	under 56 cm	56 - 58	58 - 61	61 - 63	over 63
6 months					
Mass	under 5.8 kg	5.8 - 6.8	6.8 - 7.9	7.9 - 9.1	over 9.1
Length	under 61 cm	61 - 64	64 - 67	67 - 69	over 69
9 months					
Mass	under 6.8 kg	6.8 - 8.1	8.1 - 9.4	9.4 - 11.0	over 11.0
Length	under 65 cm	65 - 69	69 - 72	72 - 74	over 74
1 year					
Mass	under 7.6 kg	7.6 - 9.0	9.0 - 10.4	10.4 - 12.3	over 12.3
Length	under 69 cm	69 - 72	72 - 76	76 - 79	over 79

64 Here is a mass-length chart for male babies from birth to one year. Masses are given in kilograms and lengths are given in centimetres.

	very low	low	Boys average	high	very high
Birth					
Mass	under 1.7 kg	1.7 - 3.2	3.2 - 3.7	3.7 - 4.5	over 4.5
Length	under 46 cm	46 - 50	50 - 52	52 - 55	over 55
3 months					
Mass	under 4.8 kg	4.8 - 5.3	5.3 - 6.1	6.1 - 7.5	over 7.5
Length	under 57 cm	57 - 59	59 - 62	62 - 64	over 64
6 months					
Mass	under 6.4 kg	6.4 - 7.0	7.0 - 8.2	8.2 - 9.4	over 9.4
Length	under 63 cm	63 - 65	65 - 68	68 - 70	over 70
9 months					
Mass	under 7.5 kg	7.5 - 8.5	8.5 - 9.8	9.8 - 11.1	over 11.1
Length	under 68 cm	68 - 70	70 - 73	73 - 76	over 76
1 year					
Mass	under 8.4 kg	8.4 - 9.4	9.4 - 10.8	10.8 - 12.4	over 12.4
Length	under 71 cm	71 - 74	74 - 77	77 - 80	over 80

On pages 66 through 69 are some interesting age-height and age-mass graphs. Of course, mass or height varies considerably within each age bracket, and so we have included three curves for each graph: lower, mean (or average) and upper.

To read the graphs, find the age at the bottom of the page, follow the vertical line up to the curves, then follow the horizontal line from lower, mean and upper curve to read the mass or height at the left. An example is given for each graph.

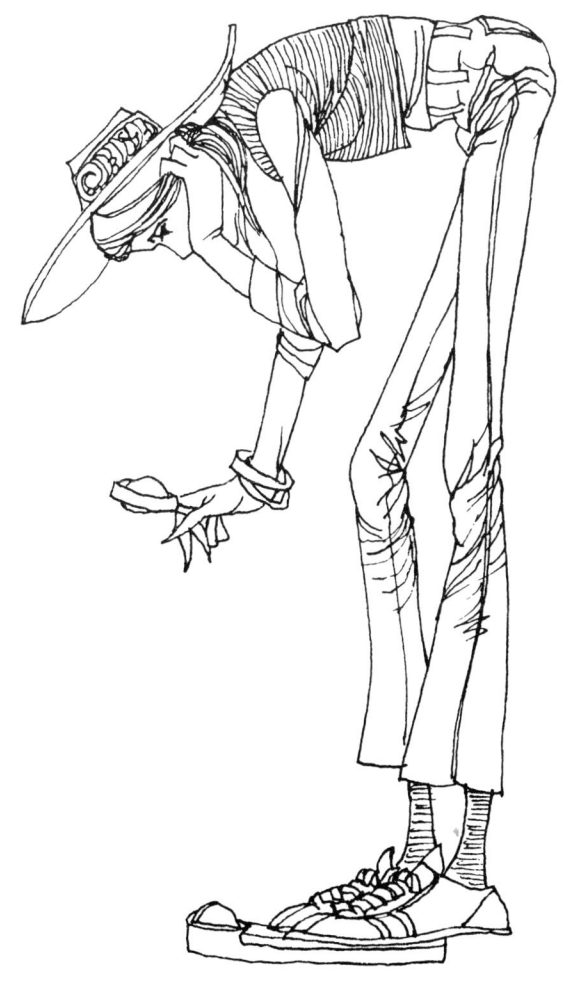

66 Age-Height Graph
Boys 1 to 18 Years

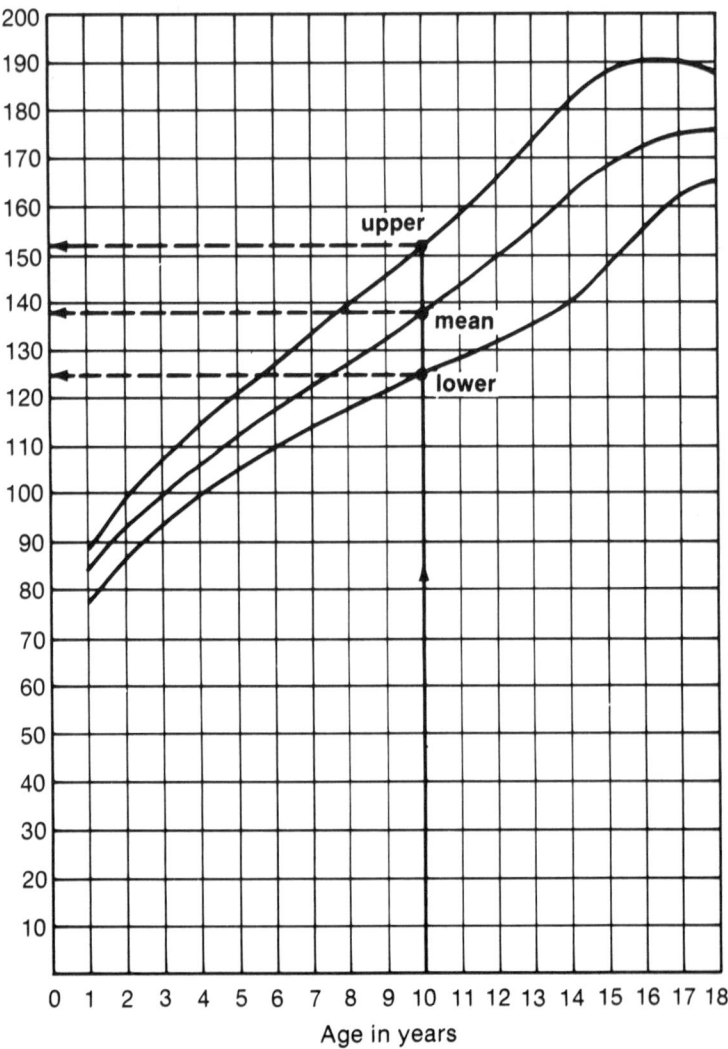

Example At age ten, a boy's height would probably be between 125 and 152 cm. The mean or average height is about 138 cm.

Age-Mass Graph
Boys 1 to 18 Years

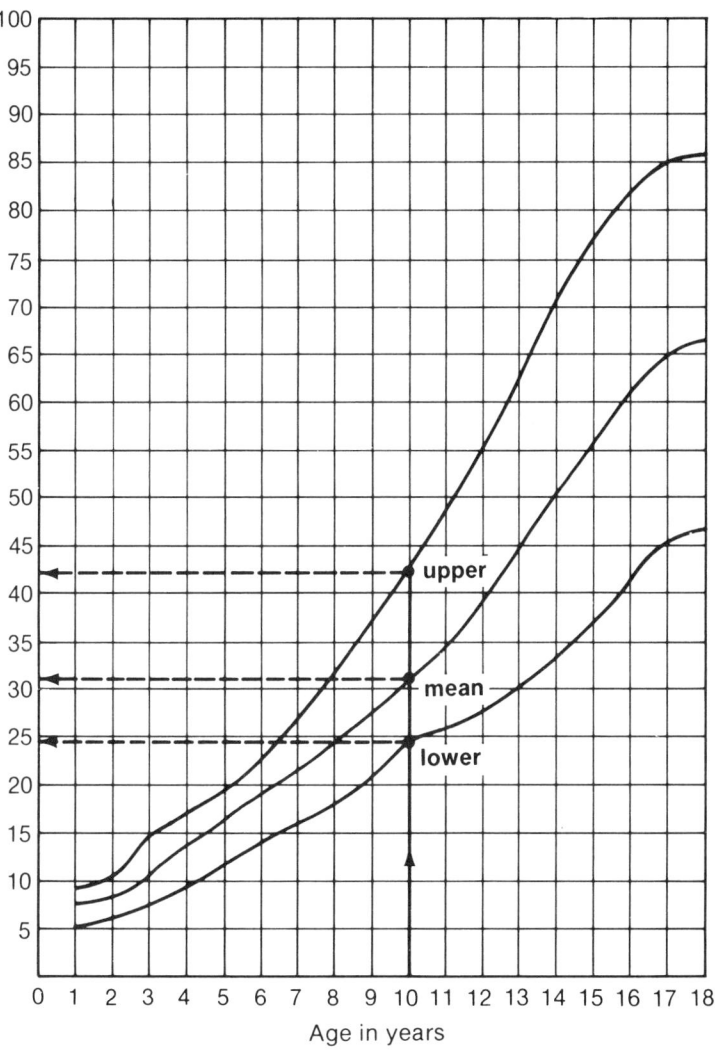

Example At age ten, a boy's mass would probably be between 24.5 and 42 kg. The average mass is about 31 kg.

68 *Age-Height Graph Girls 1 to 18 Years*

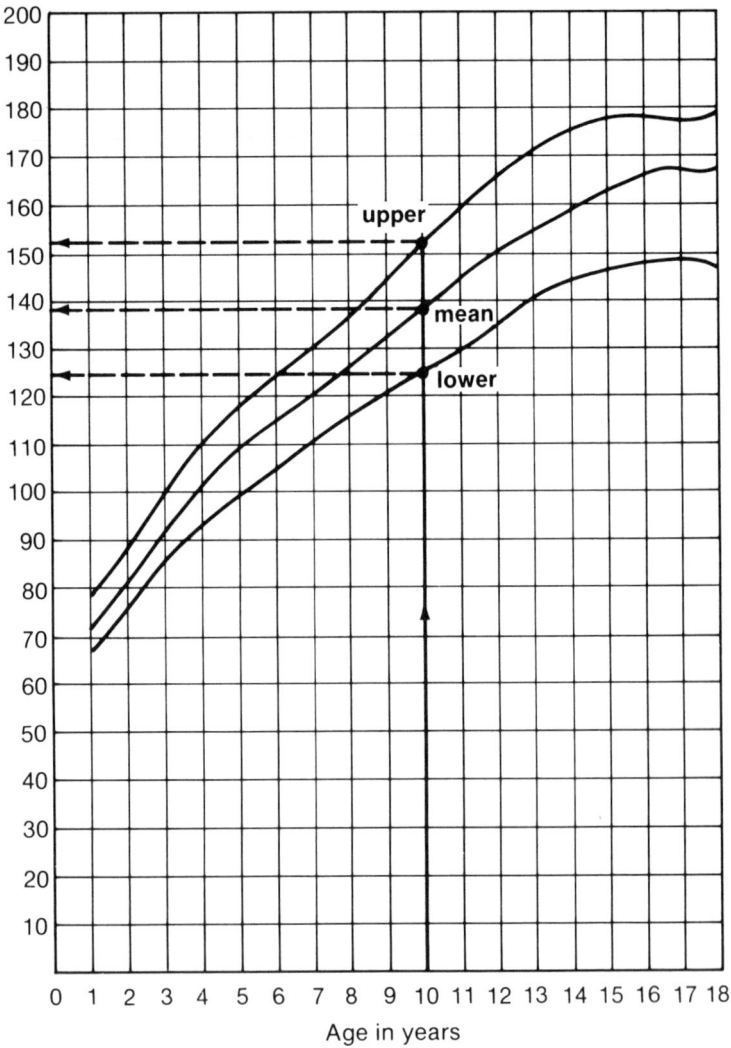

Example At age ten, a girl's height would probably be between 125 and 152 cm. The average height is about 138 cm.

Age-Mass Graph
Girls 1 to 18 Years

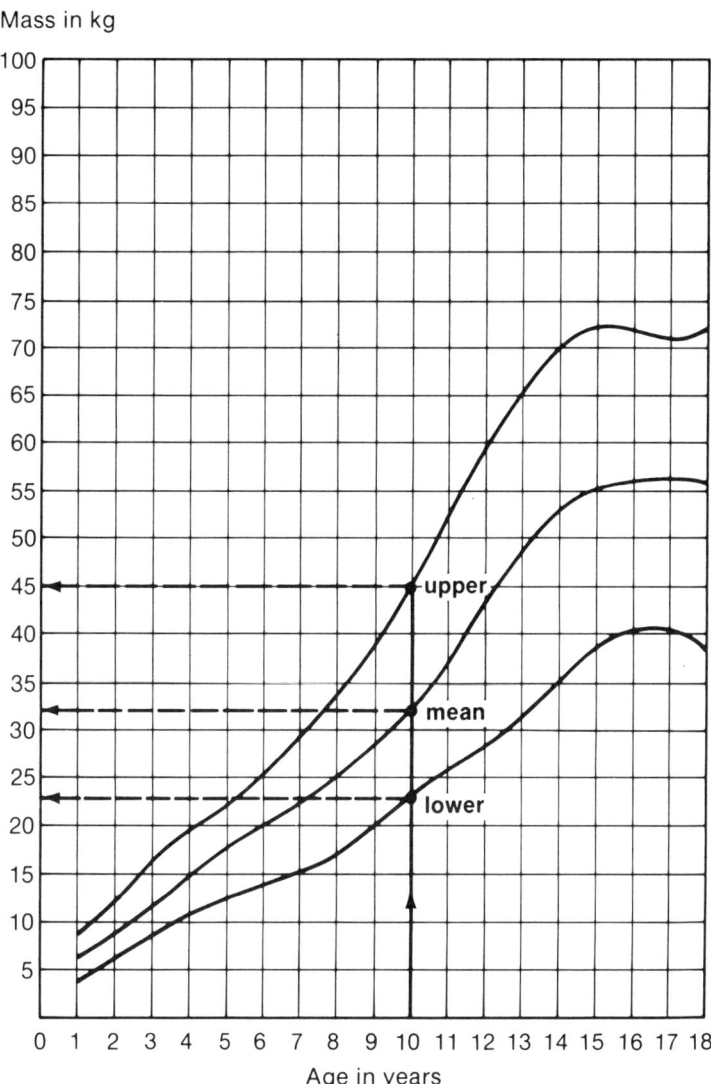

Example At age ten, a girl's mass would probably be between 23 and 45 kg. The average mass is about 32 kg.

Conversion Tables

We hope that you will make very little use of this section. Even though you have been using inches, feet, yards, pounds and quarts for many years, a change to metric is not difficult if you *think metric*.

U.S. to Metric (approximate equivalents)		
Measurement	*U.S.*	*Metric*
Linear	1 inch	2.54 cm
	1 foot	30.48 cm
	1 yard	91.44 cm
	1 mile	1.61 km
Mass	1 ounce	28.35 g
	1 pound	453.59 g
Capacity	1 quart	.946 l
Area	1 square inch	6.45 cm²
	1 square foot	0.09 m²
	1 square yard	0.84 m²
	1 acre	0.40 ha
Volume	1 cubic inch	16.39 cm³
	1 cubic foot	0.03 m³
	1 cubic yard	0.76 m³
Temperature	boiling point of water 212°F	100°C
	freezing point of water 32°F	0°C
	normal body temperature 98.6°F	37°C
Speed	30 mph	48 km/h
	31 mph	50 km/h
	45 mph	72 km/h
	70 mph	113 km/h

Measurement	Metric	U.S.
Linear	1 cm	0.39 inches
	1 m	39.37 inches
	1 km	0.62 miles
Mass	1 kg	2.20 pounds
	1 g	0.0022 pounds
	1 g	0.04 ounces
Capacity	1 ℓ	1.057 quarts
Area	1 cm²	0.15 square inches
	1 m²	1.19 square yards
	1 ha	2.47 acres
Volume	1 cm³	0.06 cubic inches
	1 m³	35.31 cubic feet
	1 m³	1.31 cubic yards

Metric to U.S.
(approximate equivalents)

Answers

Page 19   (a) CD — 4.5 cm
                       45 mm

              (b) EF — 6 cm
                       60 mm

              (c) GH — 0.7 cm
                       7 mm

              (d) IJ — 0.5 cm
                       5 mm

              (e) PQ — 6.2 cm
                       62 mm

              (f) RS — 7.2 cm
                       72 mm

              (g) UV — 7 cm
                       70 mm

              (h) XY — 9.3 cm
                       93 mm

              (i) ZW — 8.1 cm
                       81 mm

Page 39   (a) Area PQRS = (4 X 2) cm²
                                   = 8 cm²

              (b) Area MNOP = (3 X 4) cm²
                                   = 12 cm²

              (c) Area KLMN = (5 X 1) cm²
                                   = 5 cm²

              (d) Area ABCD = (3 X 3) cm²
                                   = 9 cm²

              (e) Area WXYZ = (4 X 1) cm²
                                   = 4 cm²

              (f) Area EFGH = (6 X 2) cm²
                                   = 12 cm²

Page 48 (a) Volume = 8 cm³
(b) Volume = 24 cm³
(c) Volume = 125 cm³

Page 57
1. The 3 tells the number of tens. The 0 tells the number of hundreds.
2. (a) 493
   (b) 5.76
   (c) 6
   (d) 0.0006
   (e) 5.764
   (f) 111
3. 342.36
4. 488.69
5. $2382.74
6. $8369.77
7. 26.10
8. 1015.868
9. 3.4
10. 8.6

Page 61
(a) 50 cm = 0.5 m
(b) 43.6 km = 43,600 m
(c) 67.8 cm = 0.678 m
(d) 42.6 m = 0.0426 km
(e) 43,658 cm = 0.43658 km
(f) 756 ml = 0.756 ℓ
(g) 0.3561 ℓ = 356.1 ml

Page 62
1. (a) 4.45 (b) 26.50 (c) 51.11 (d) 66.24
   (e) 221.06

2. (a) 3.088 (b) 11.464 (c) 4.294 (d) 114.17
   (e) 1639.089

3. (a) 14.85 (b) 374.32 (c) 19.09375
   (d) 50.700658

4. (a) 2.7 (b) 47.8 (c) 31.04